A life Of Radley coloring book

AF815574

Flora's Fashionable Friends

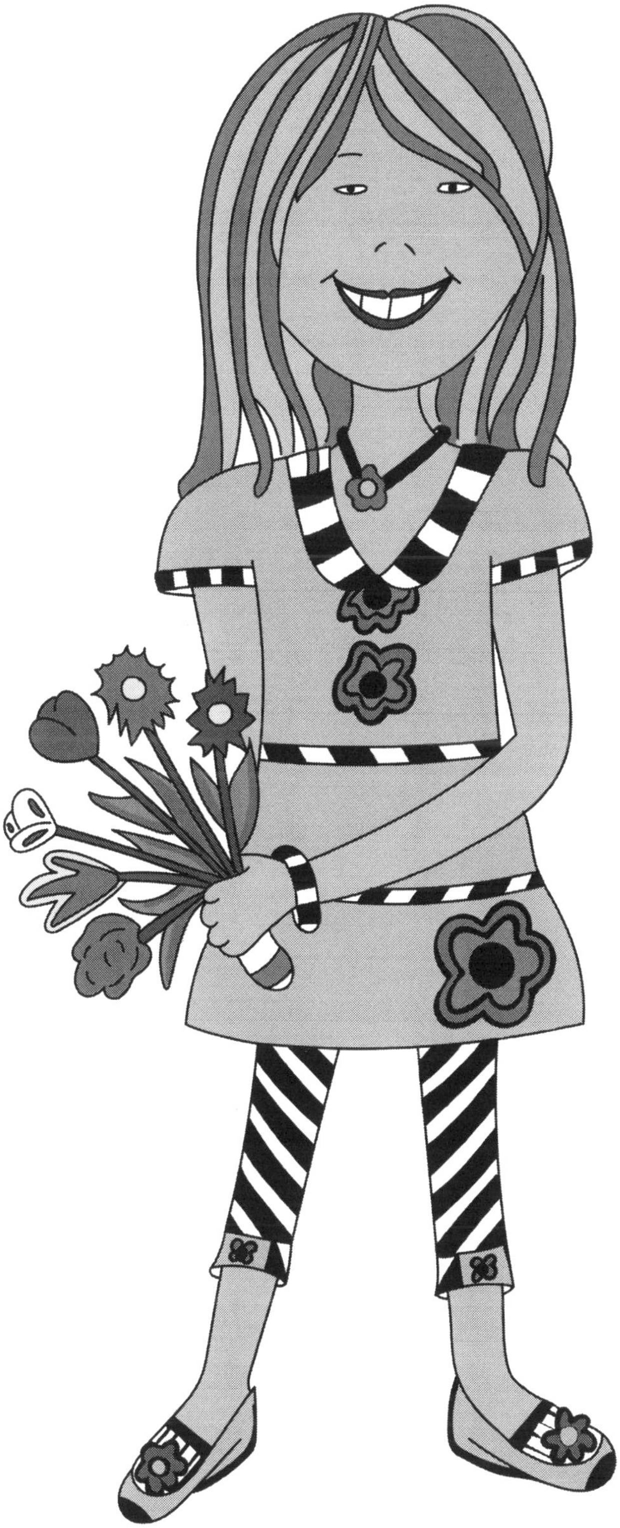

Illustrated by Beth Cope

Flora's Fashionable Friends Coloring Book

A Life of Radley Experience. Live Life Radley!

Life of Radley, Inc.
www.lifeofradley.com
info@lifeofradley.com

ISBN: 978-1-7376628-4-6

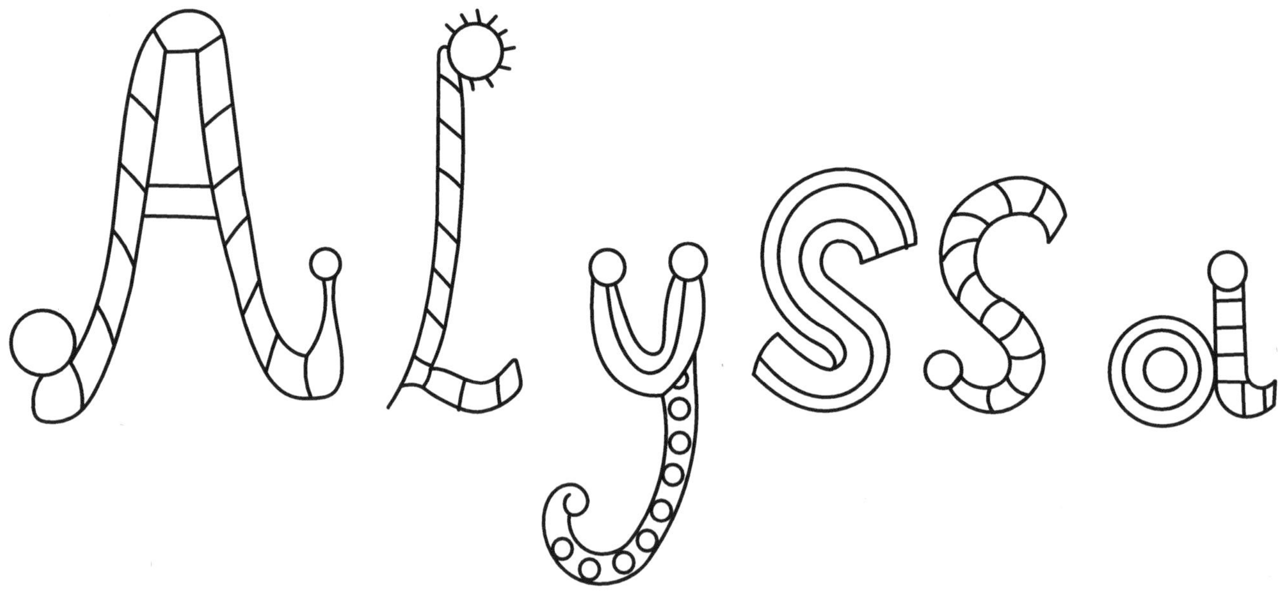
Alyssa

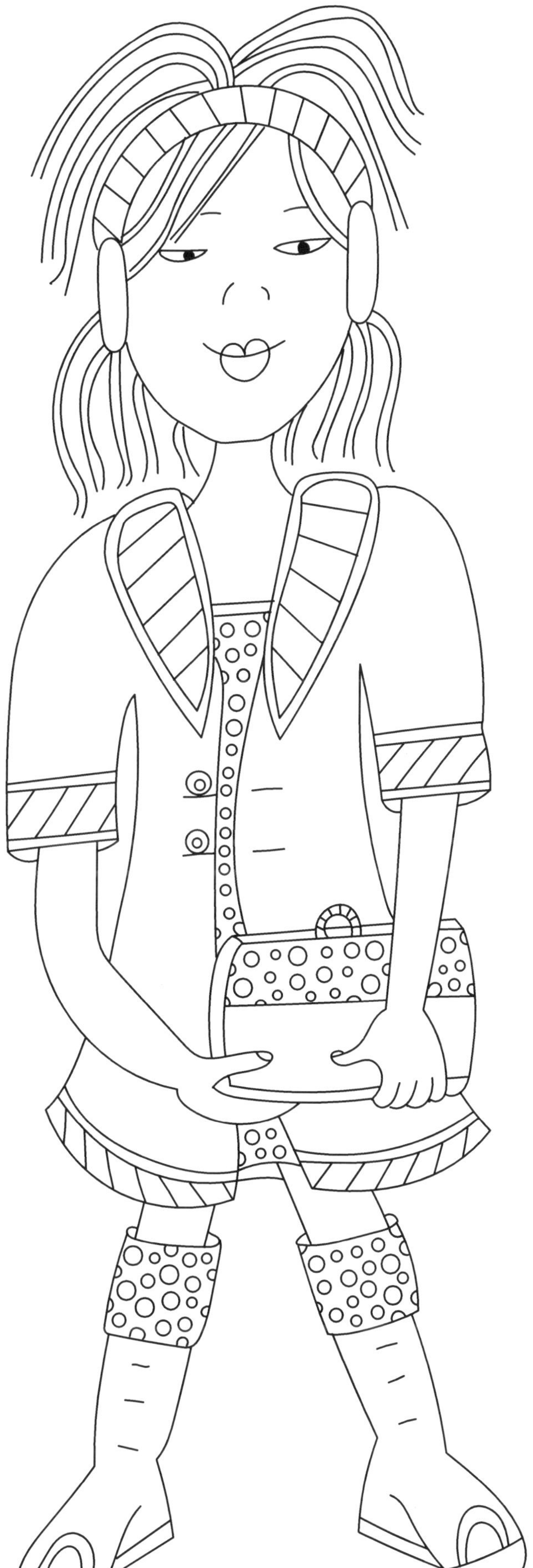

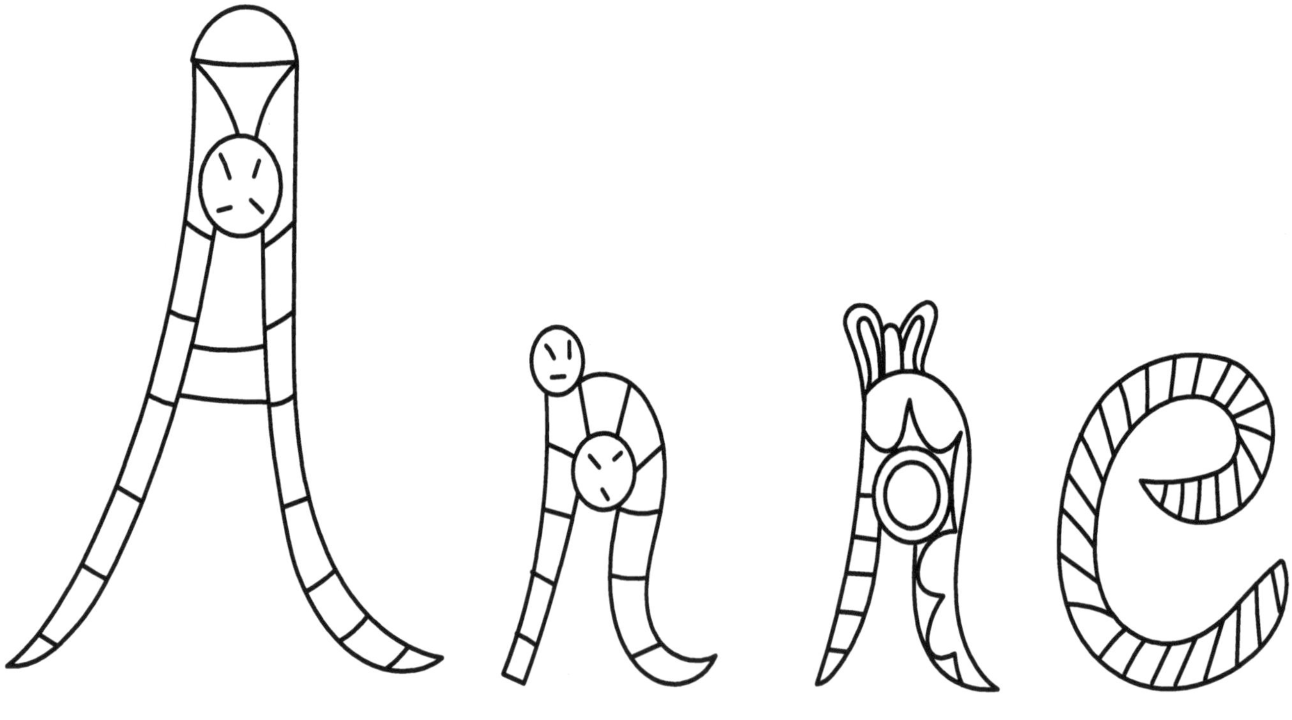
Anne

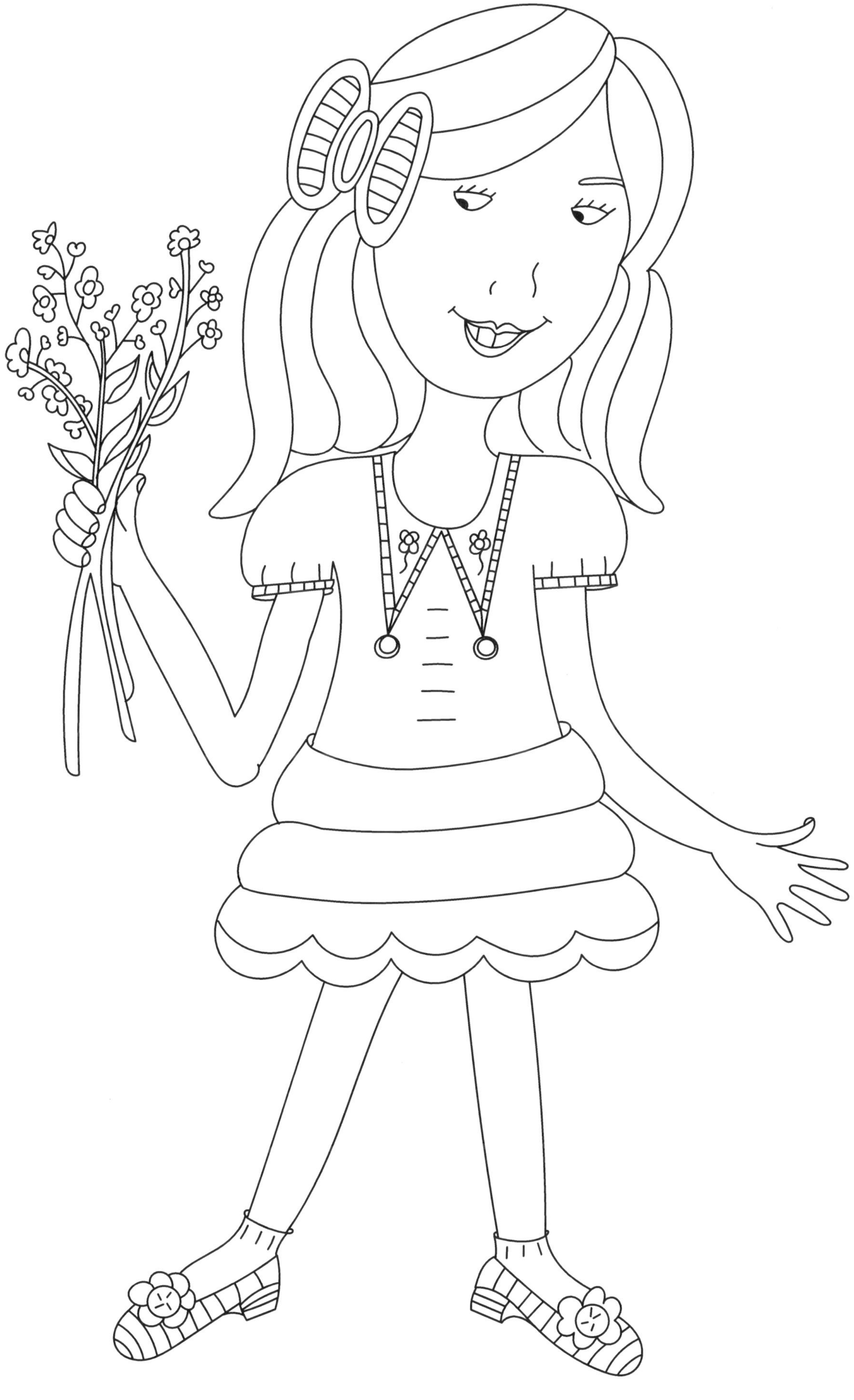

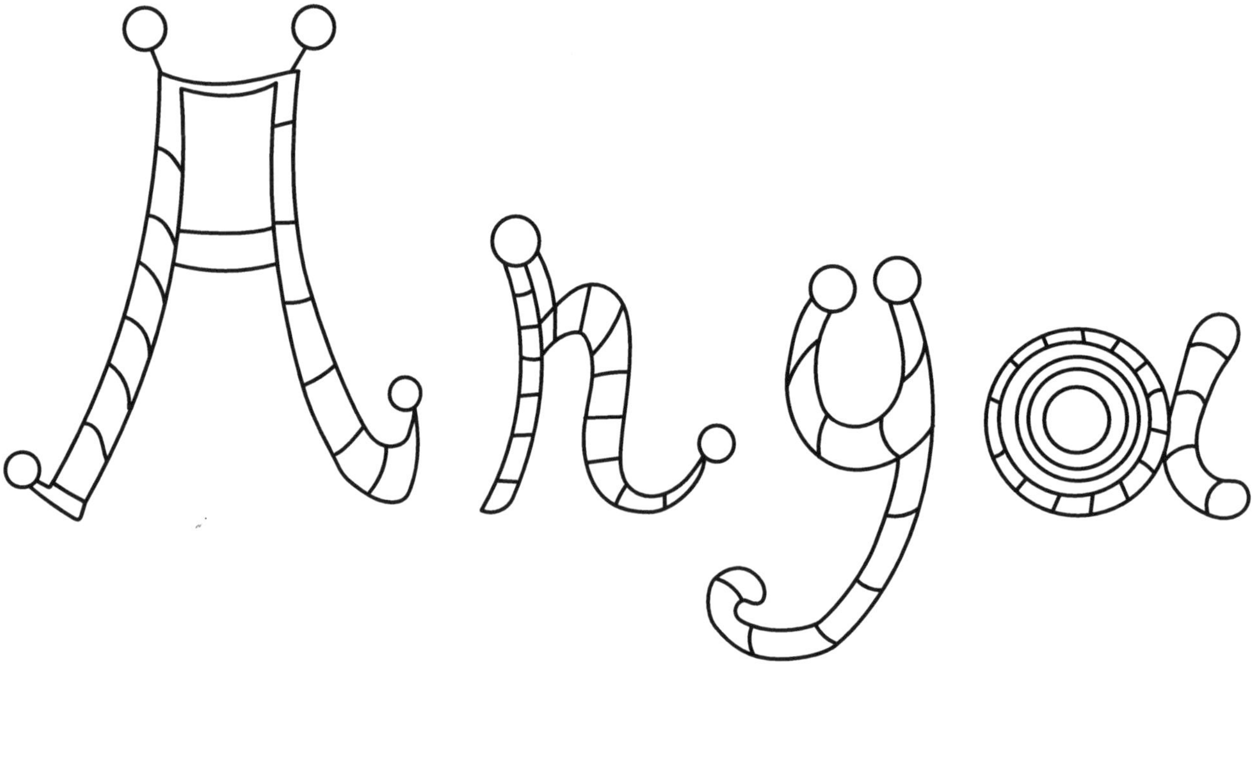
Anya

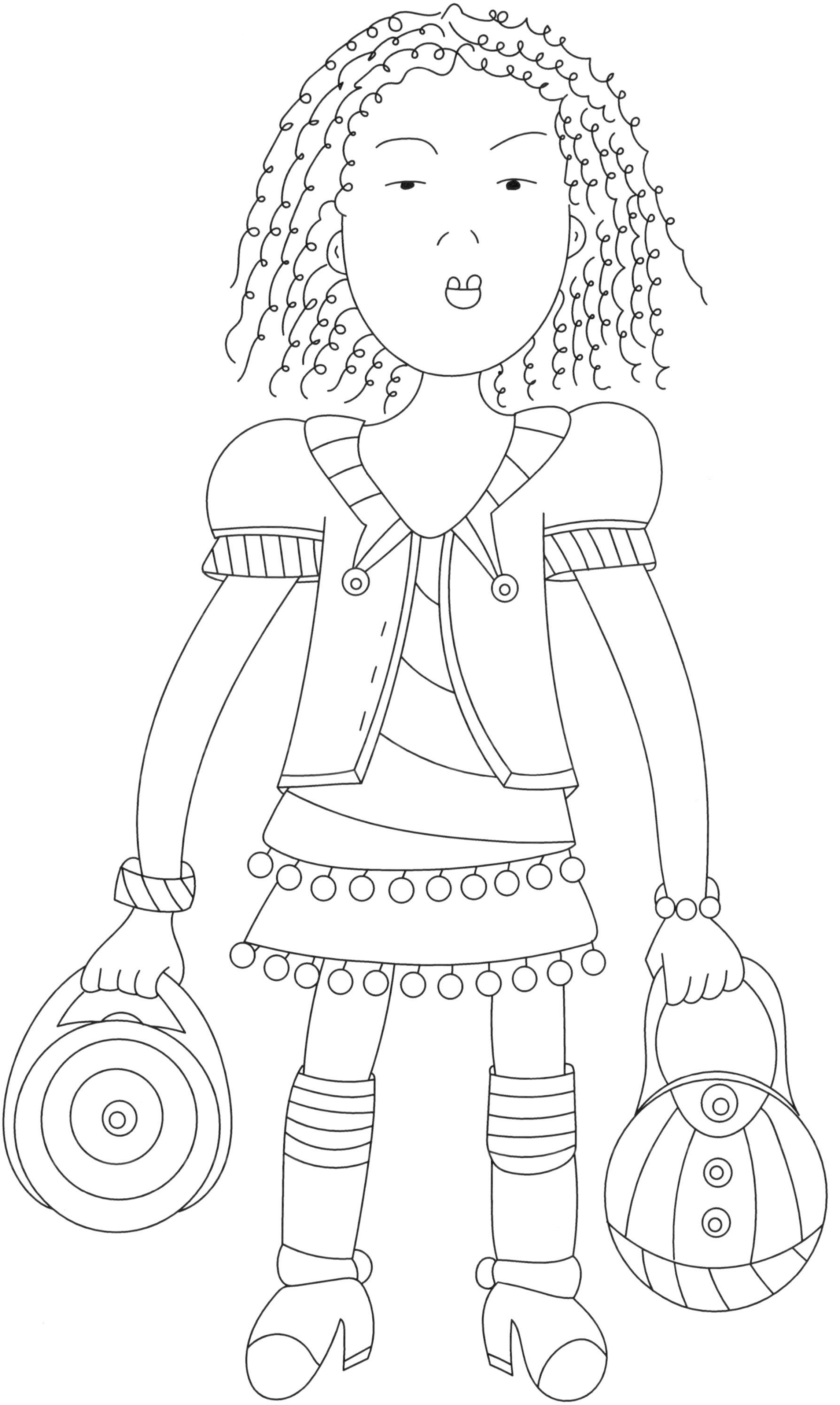

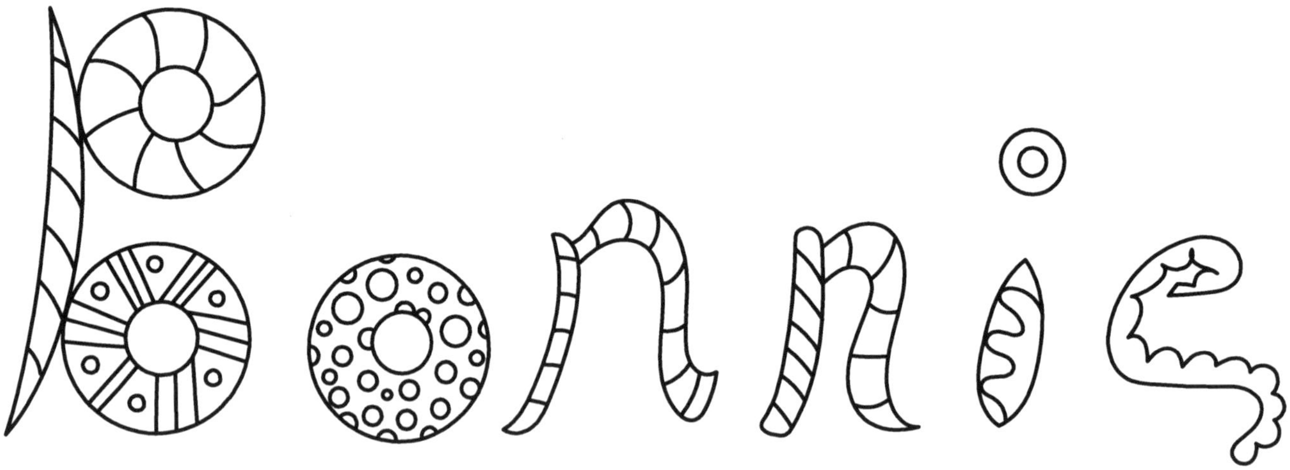
Bonnie

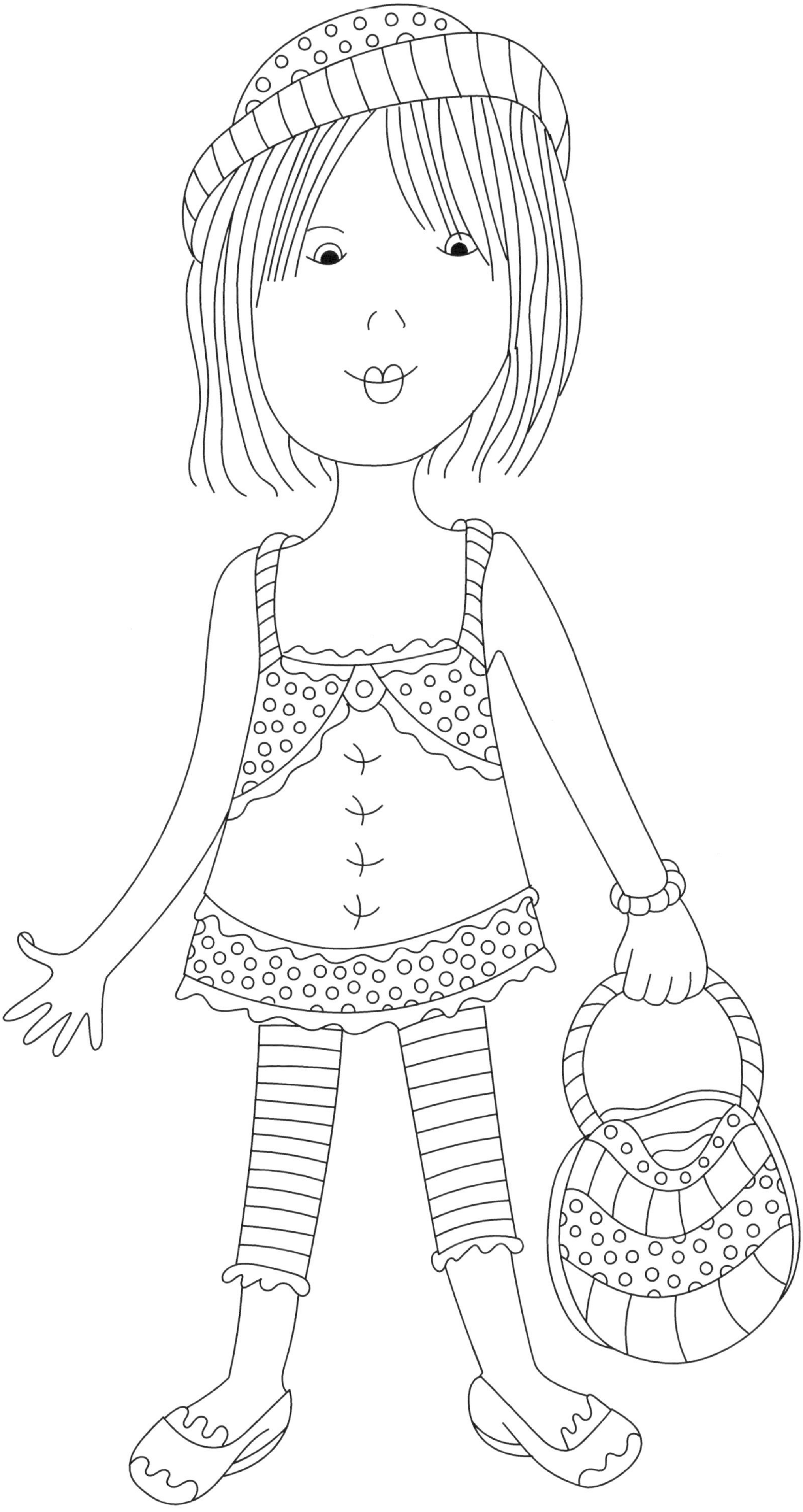

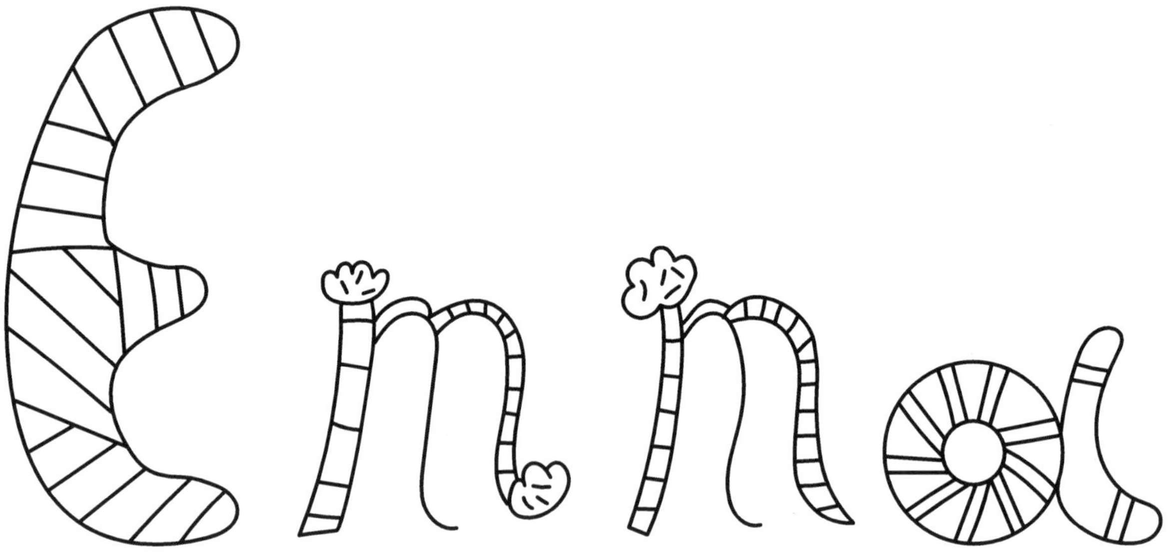
Emma

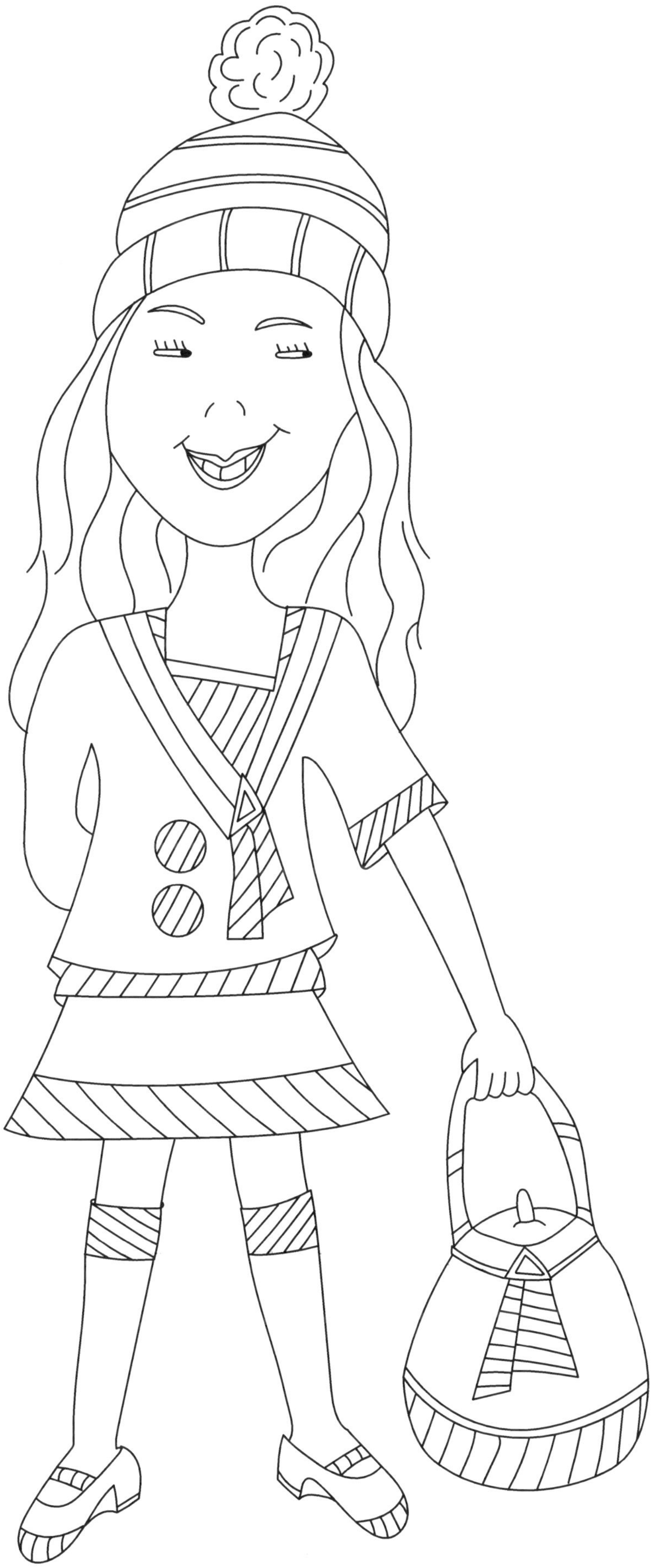

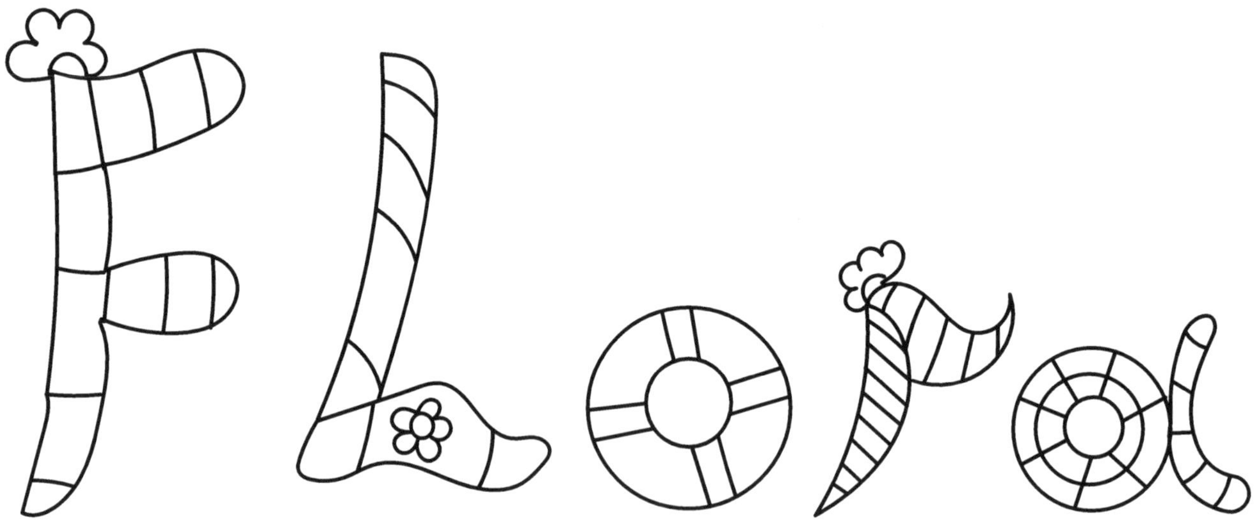
Flora

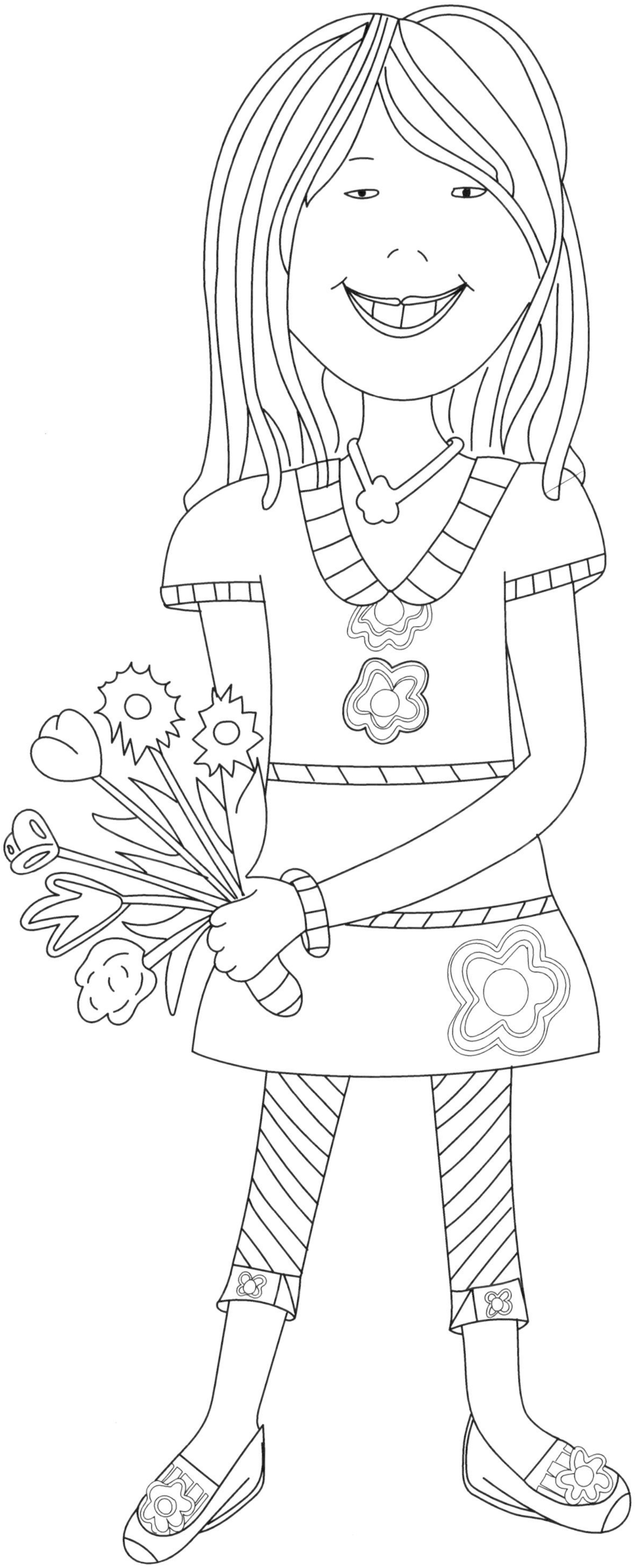

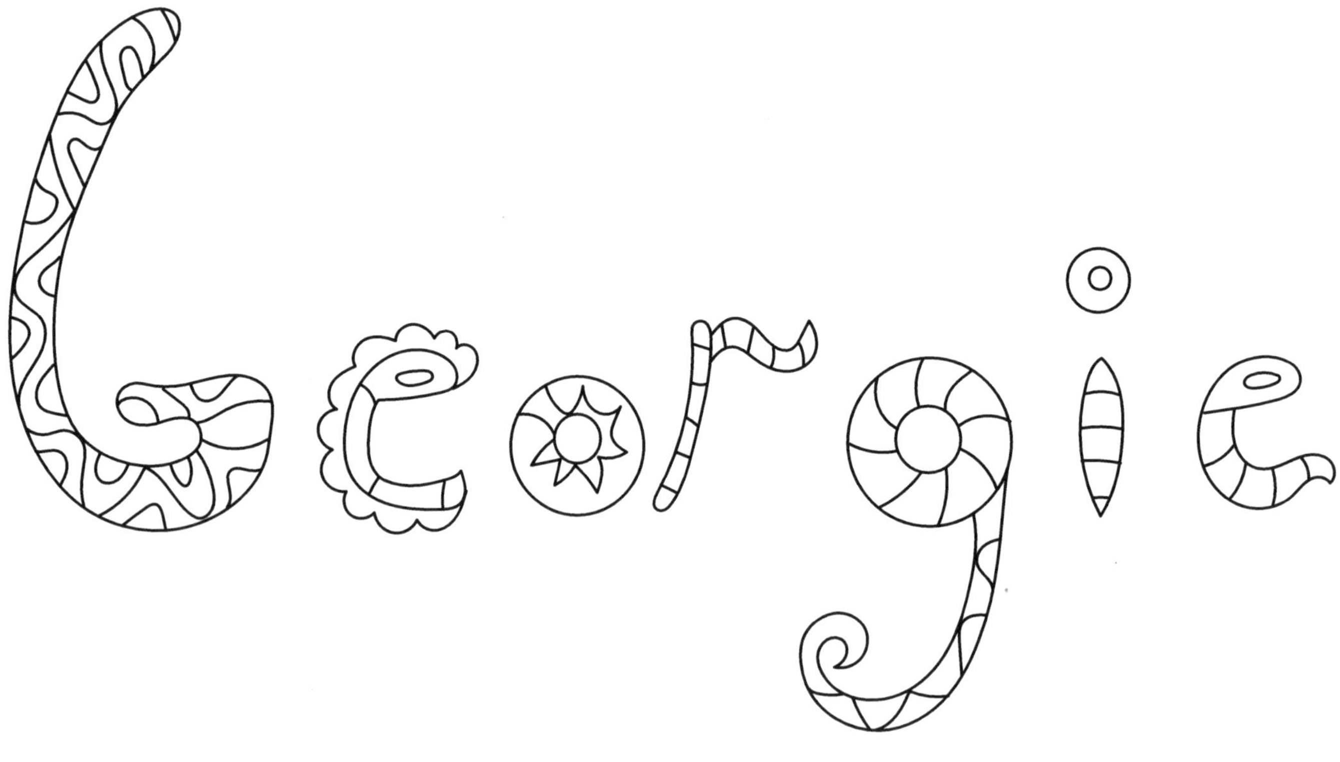
Georgie

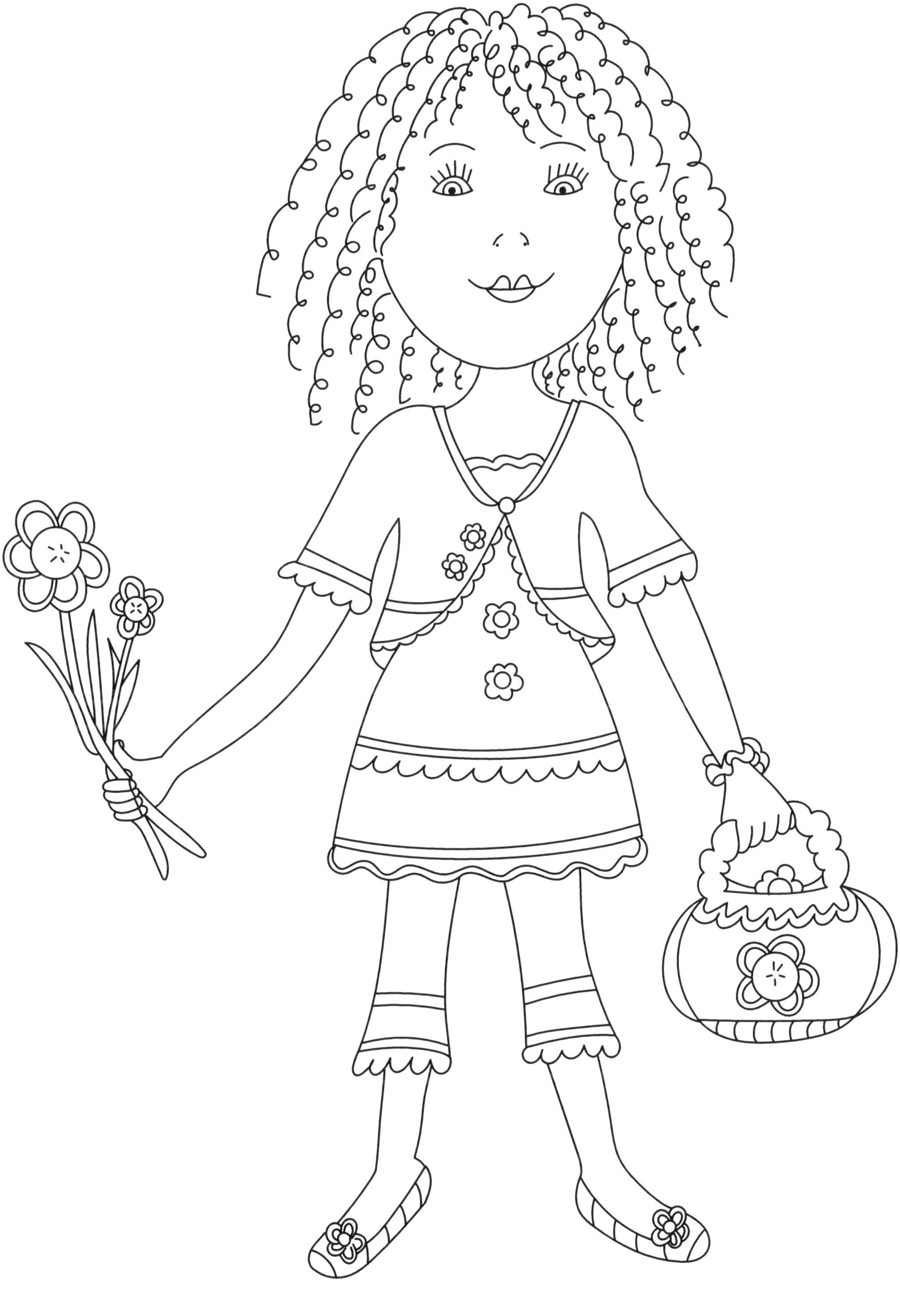

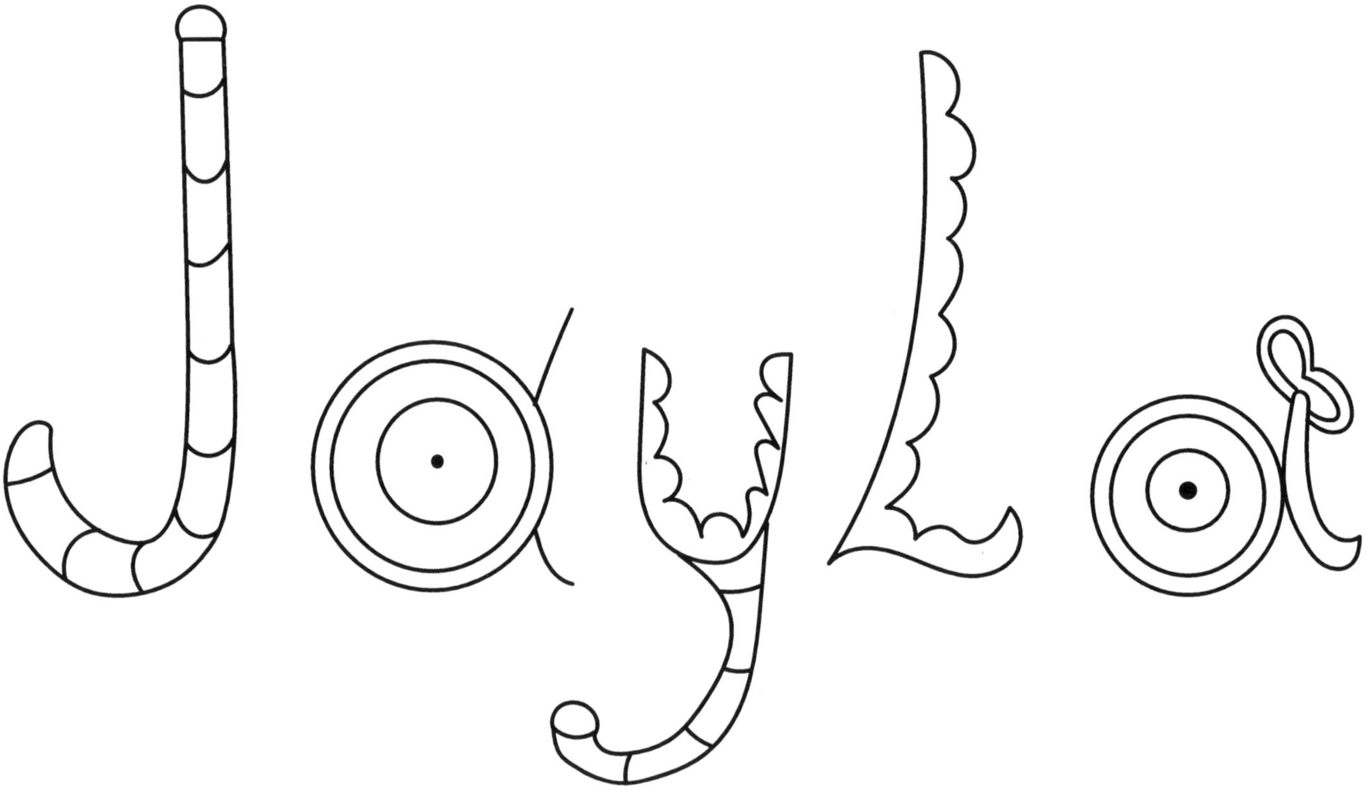
JayLa

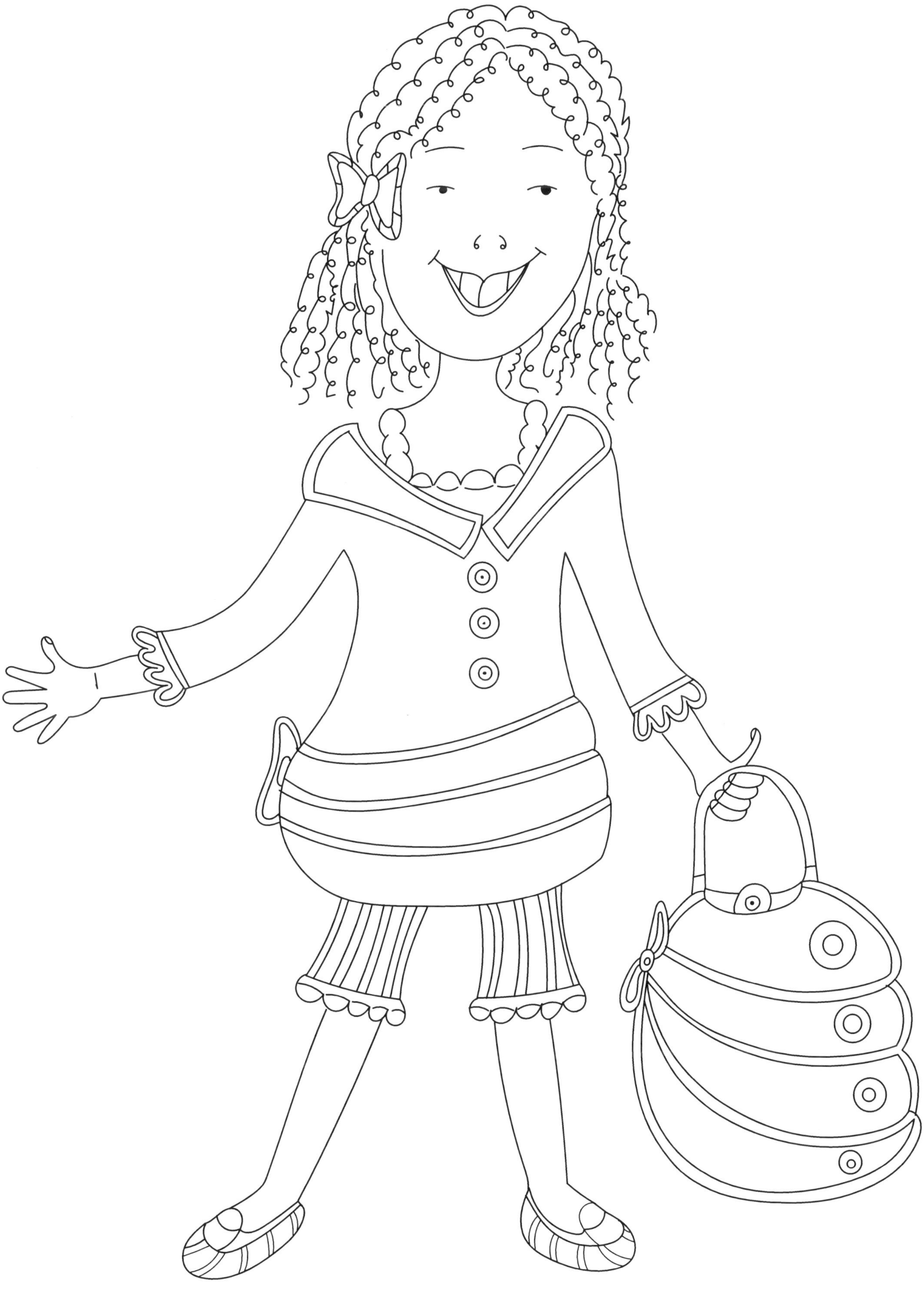

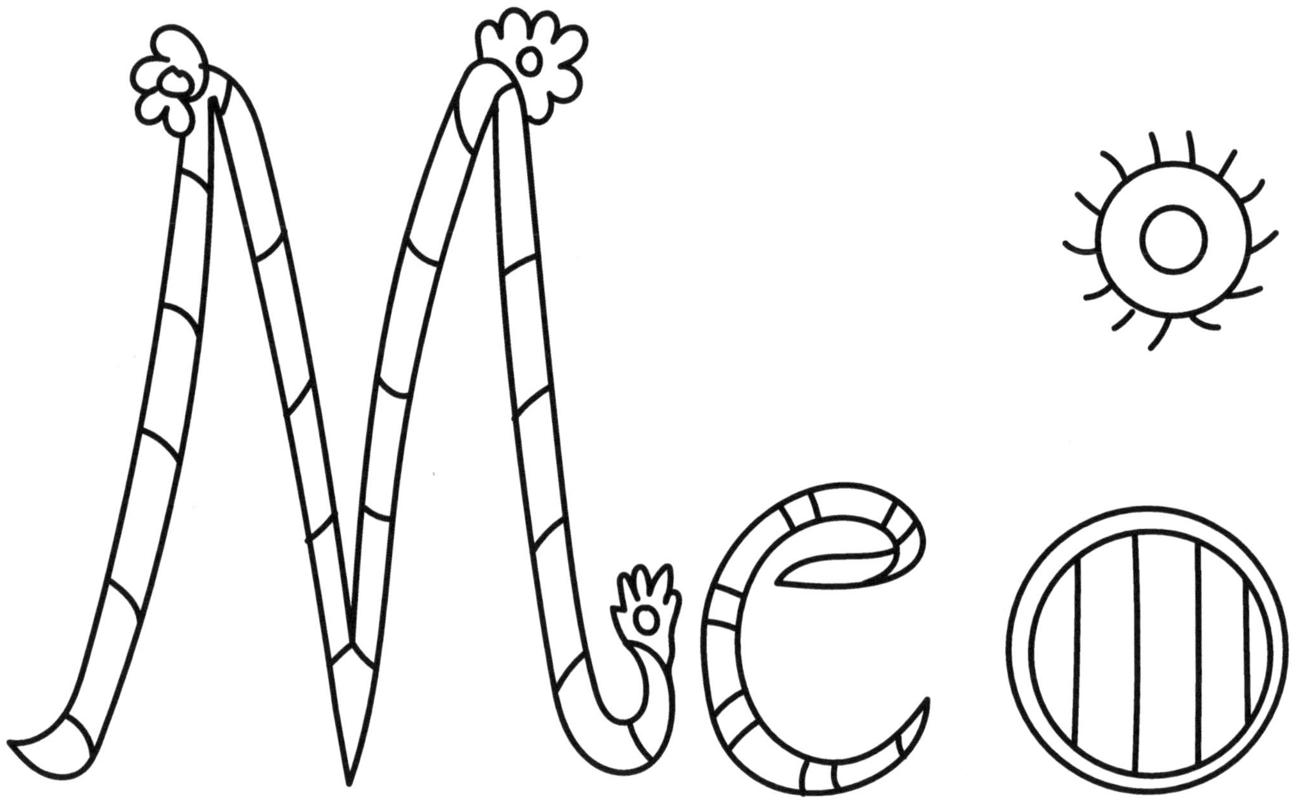
Meo

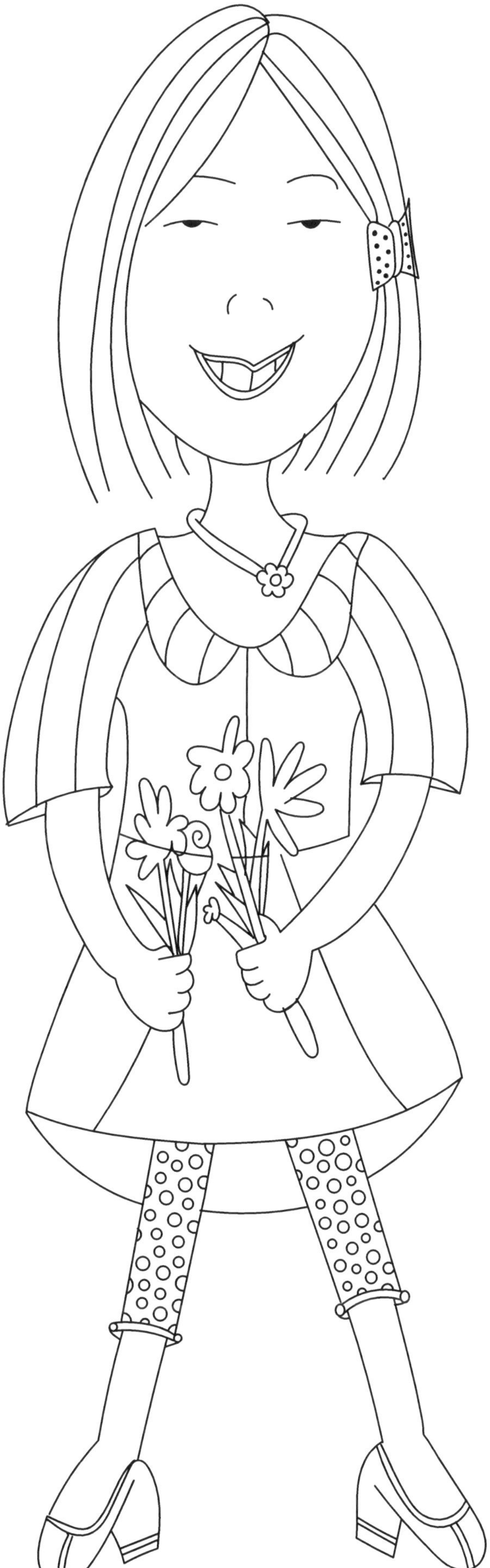

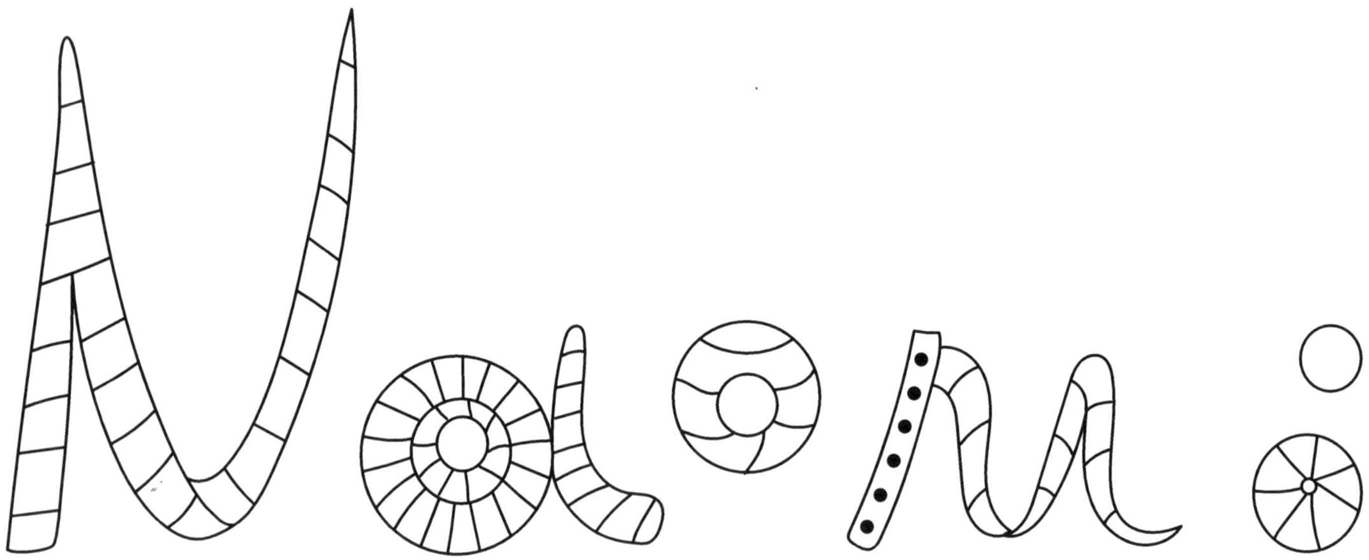
Naomi

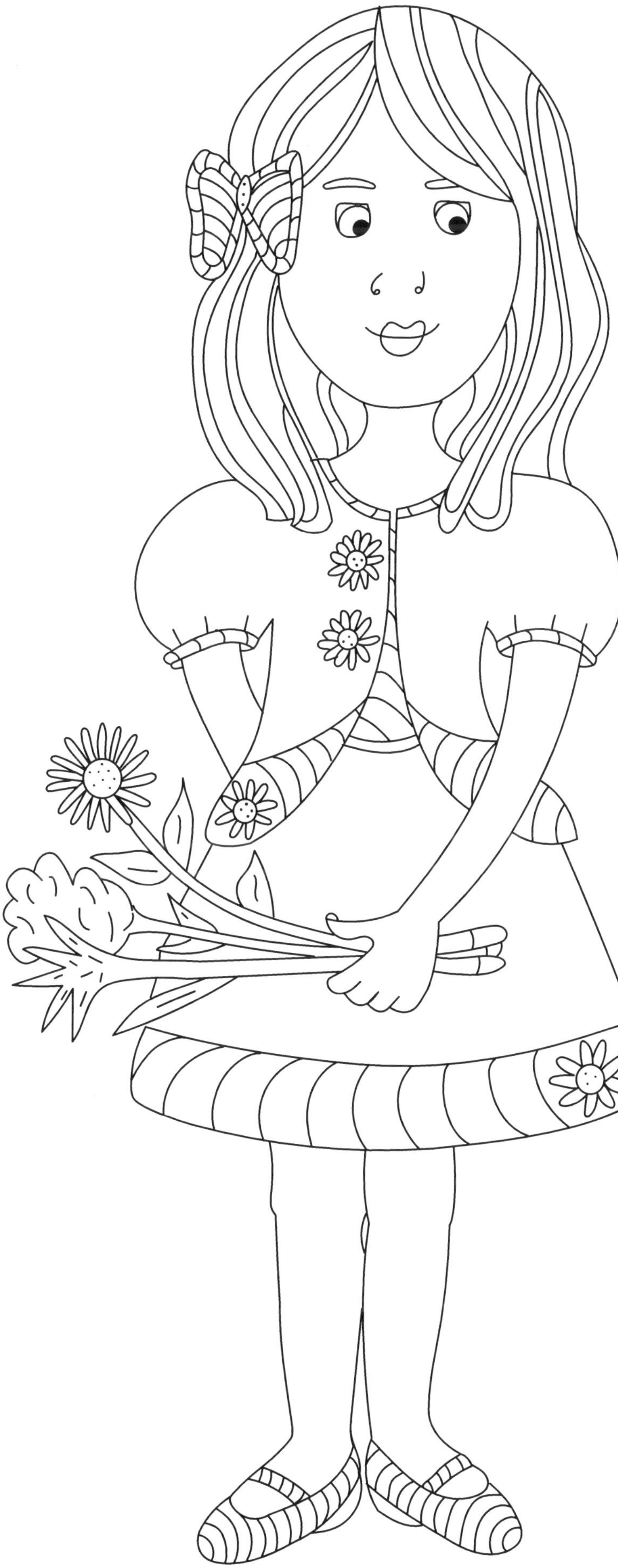

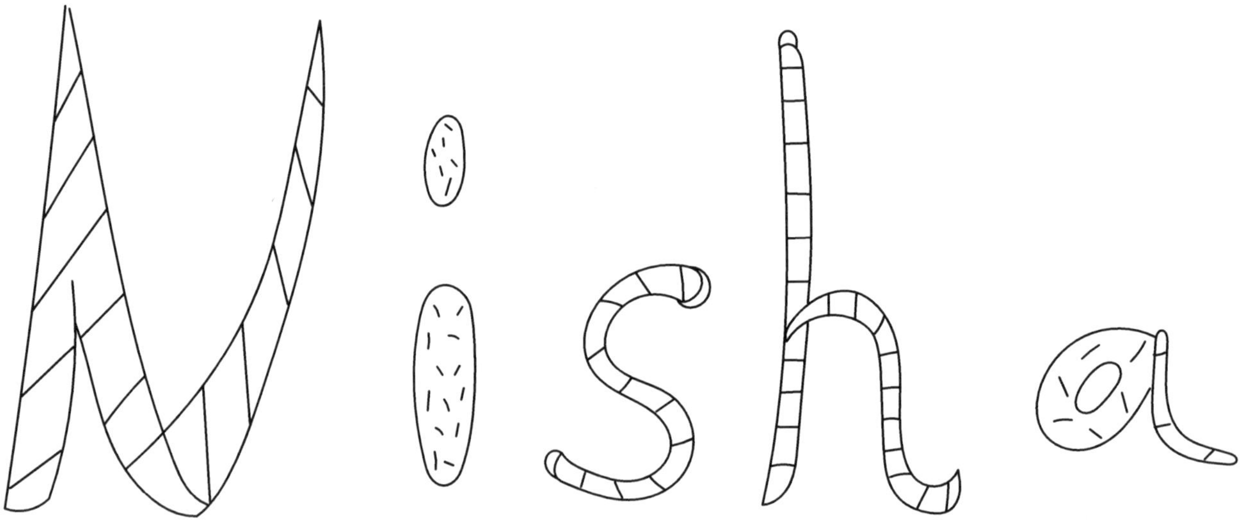
Nisha

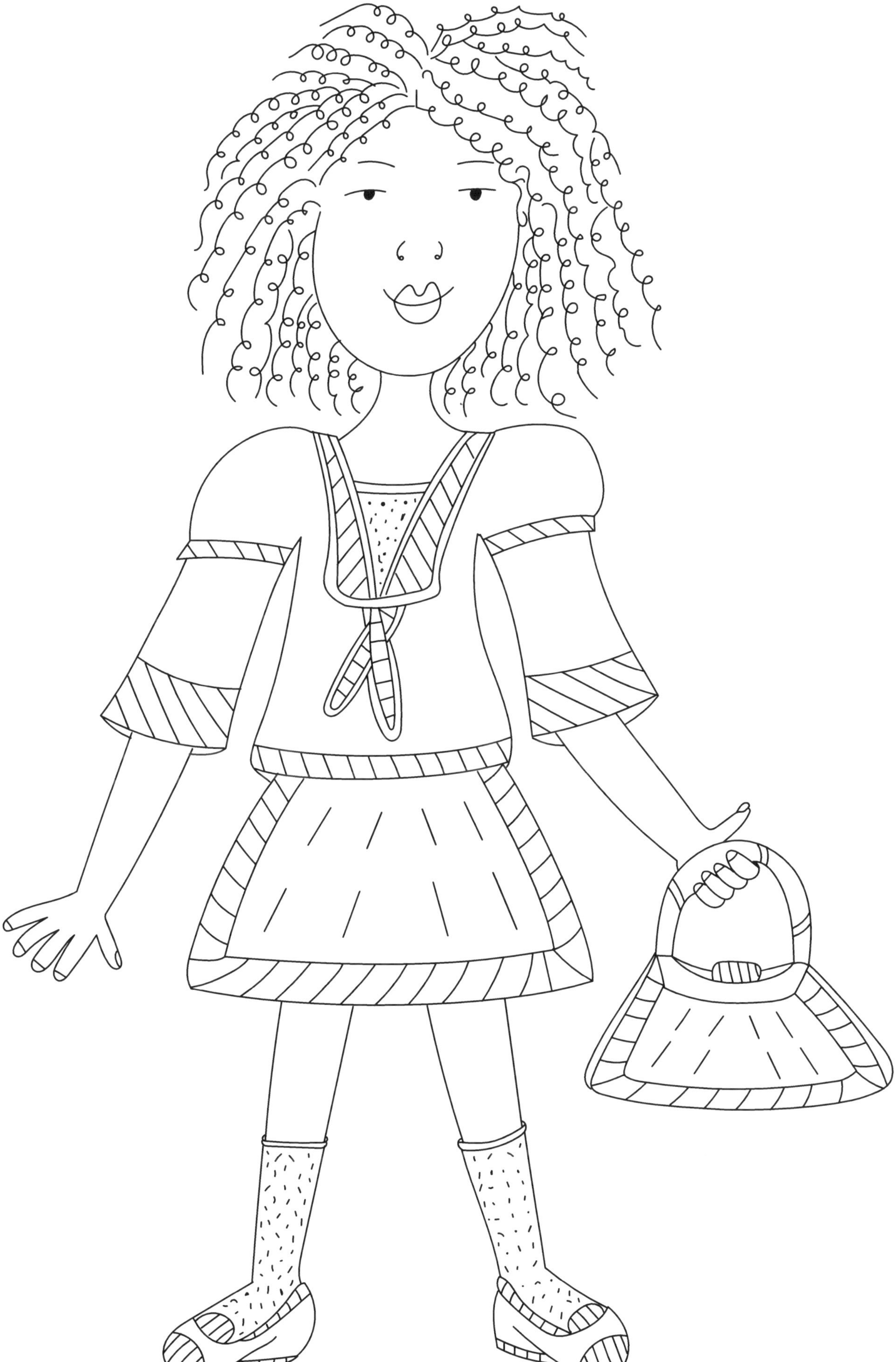

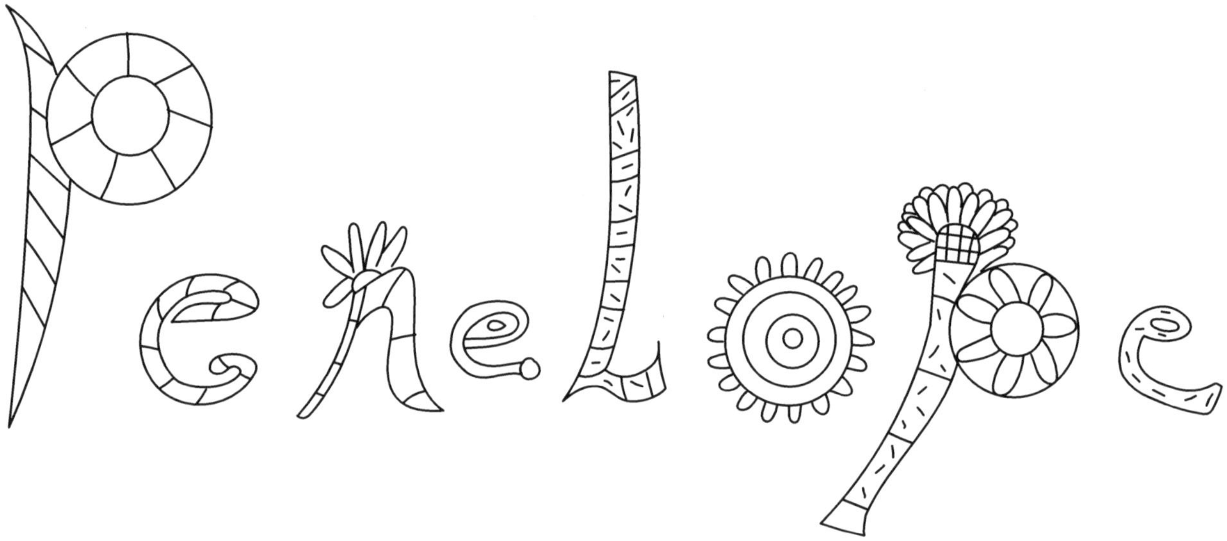
Penelope

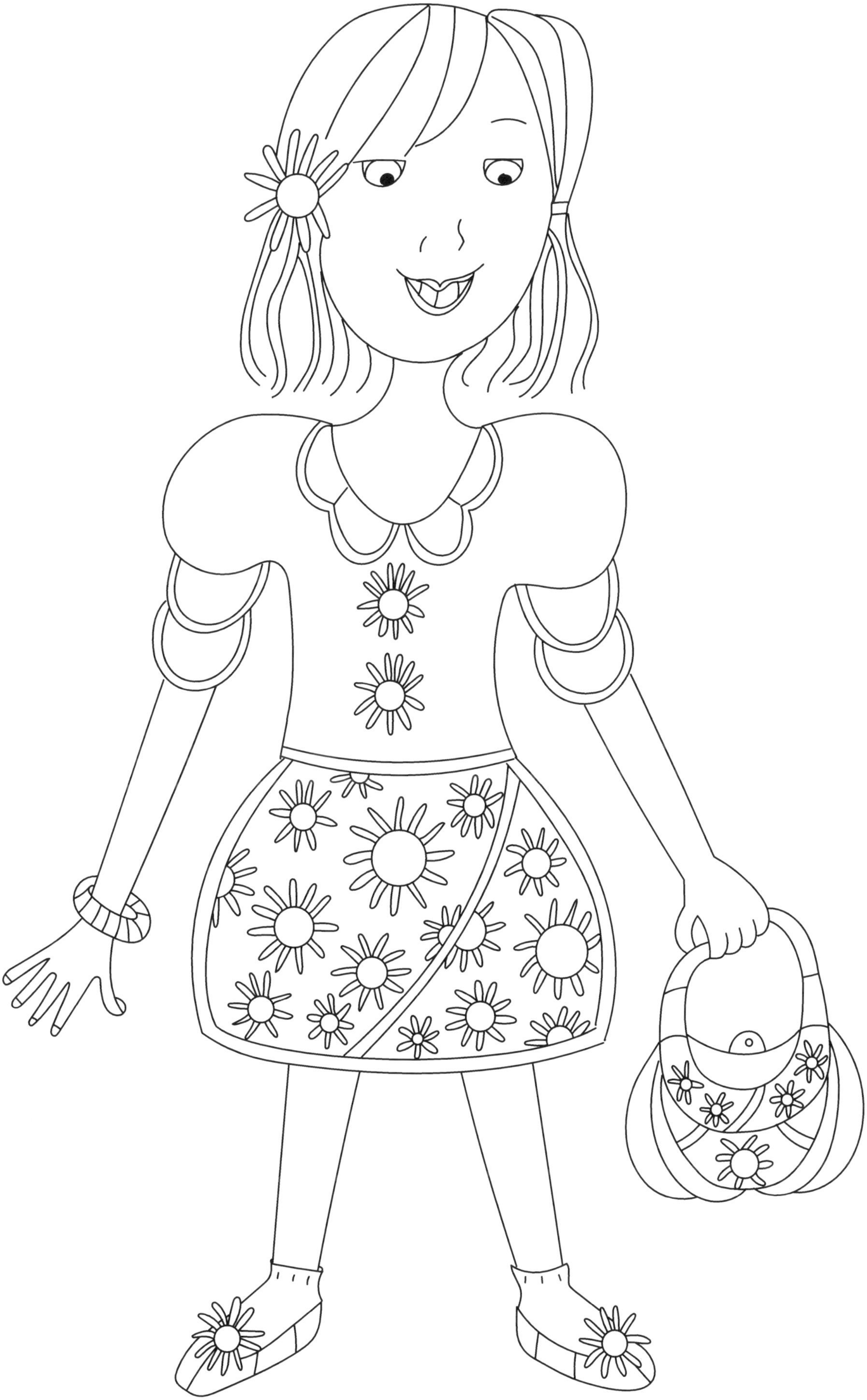

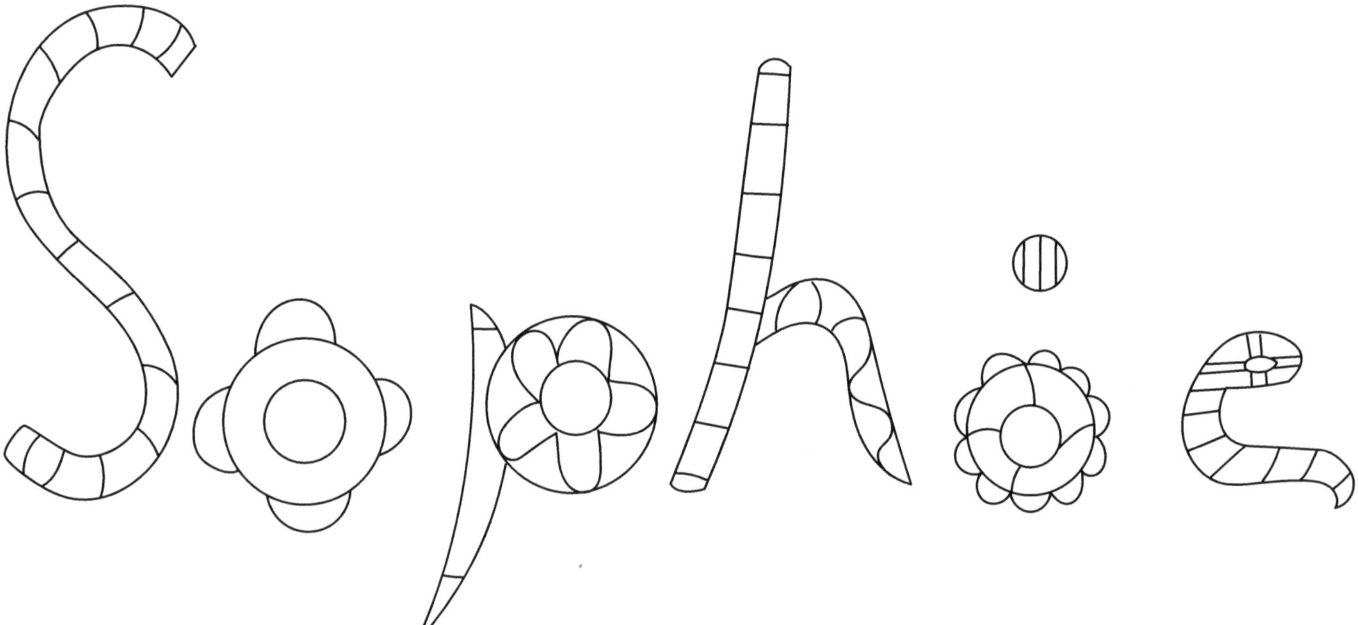
Sophie

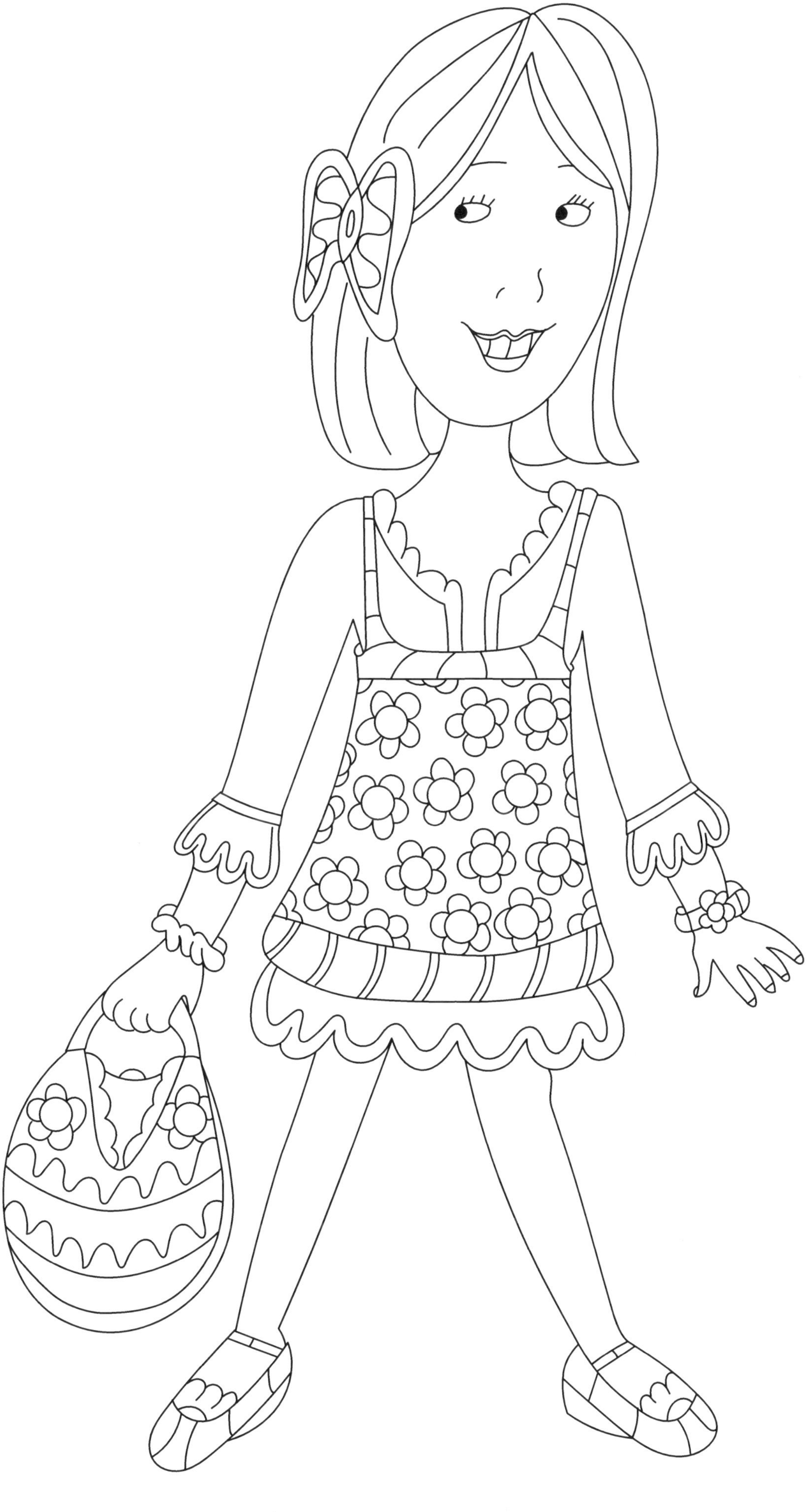

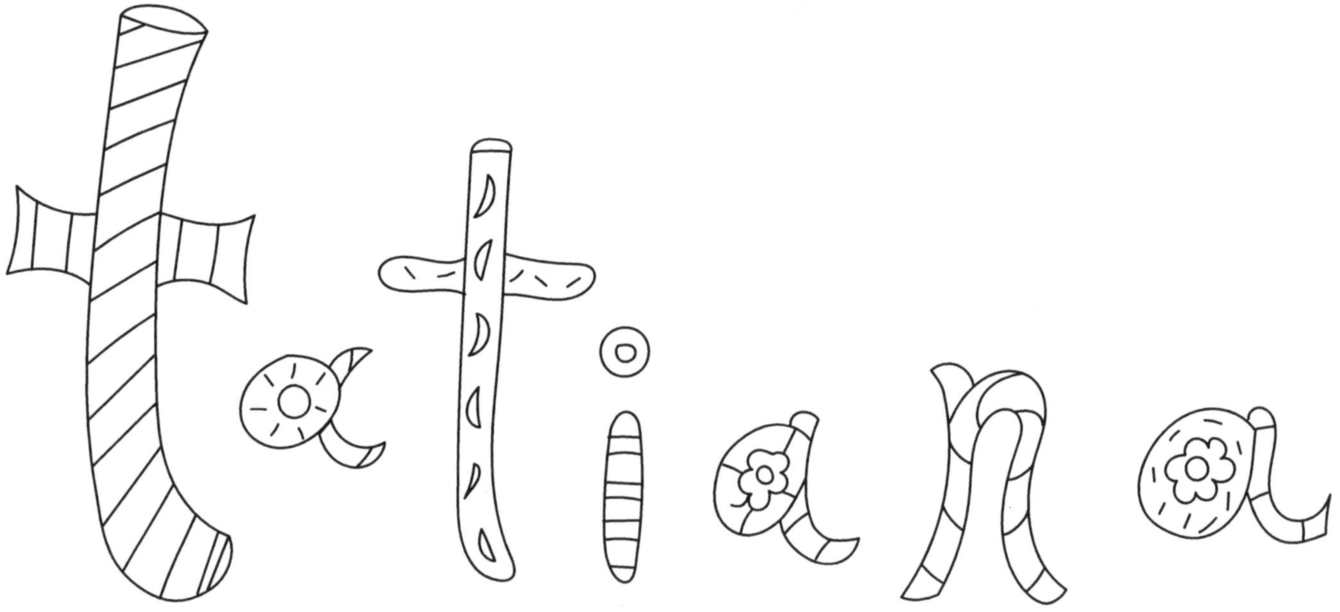
tatiana

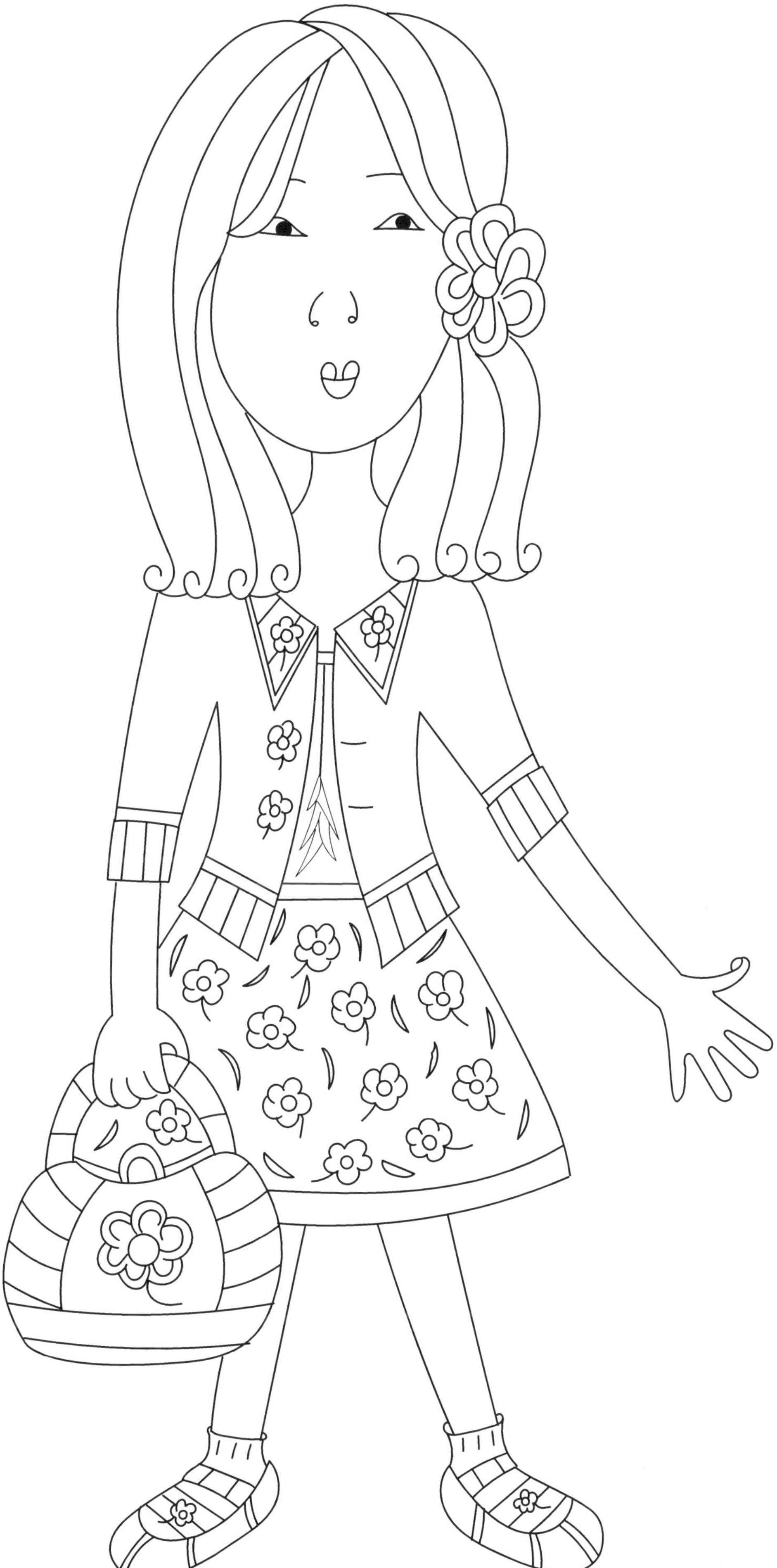

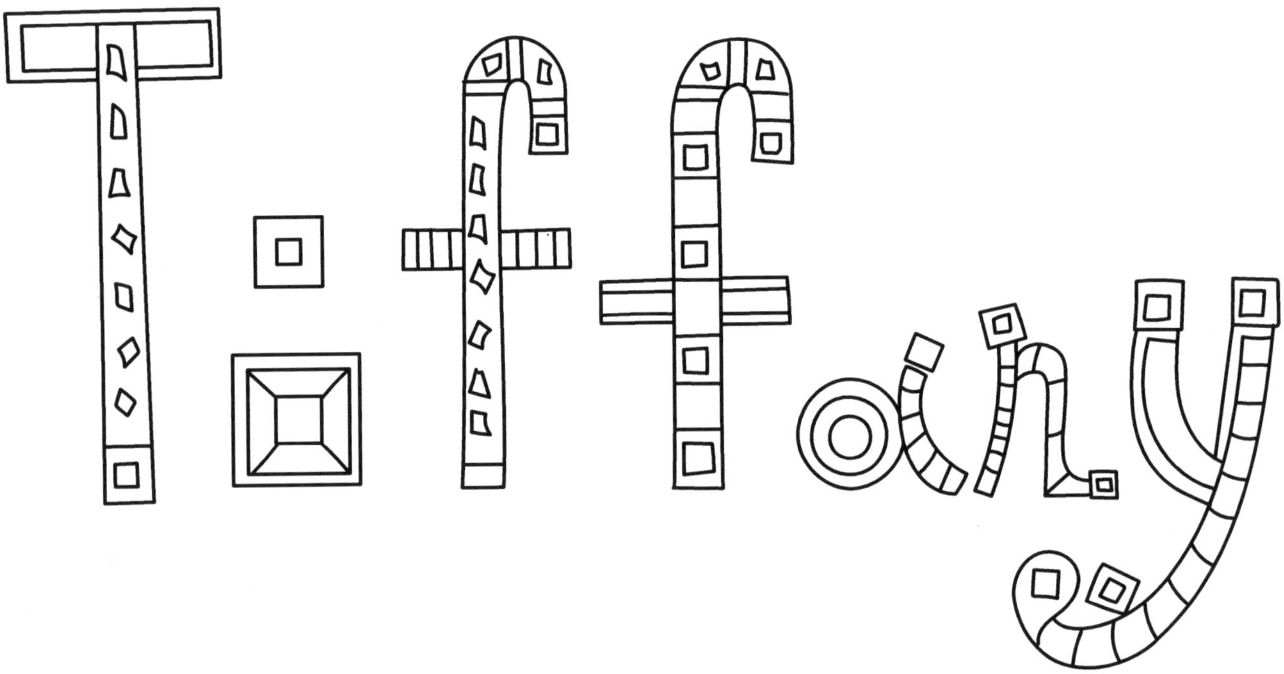
Tiffany

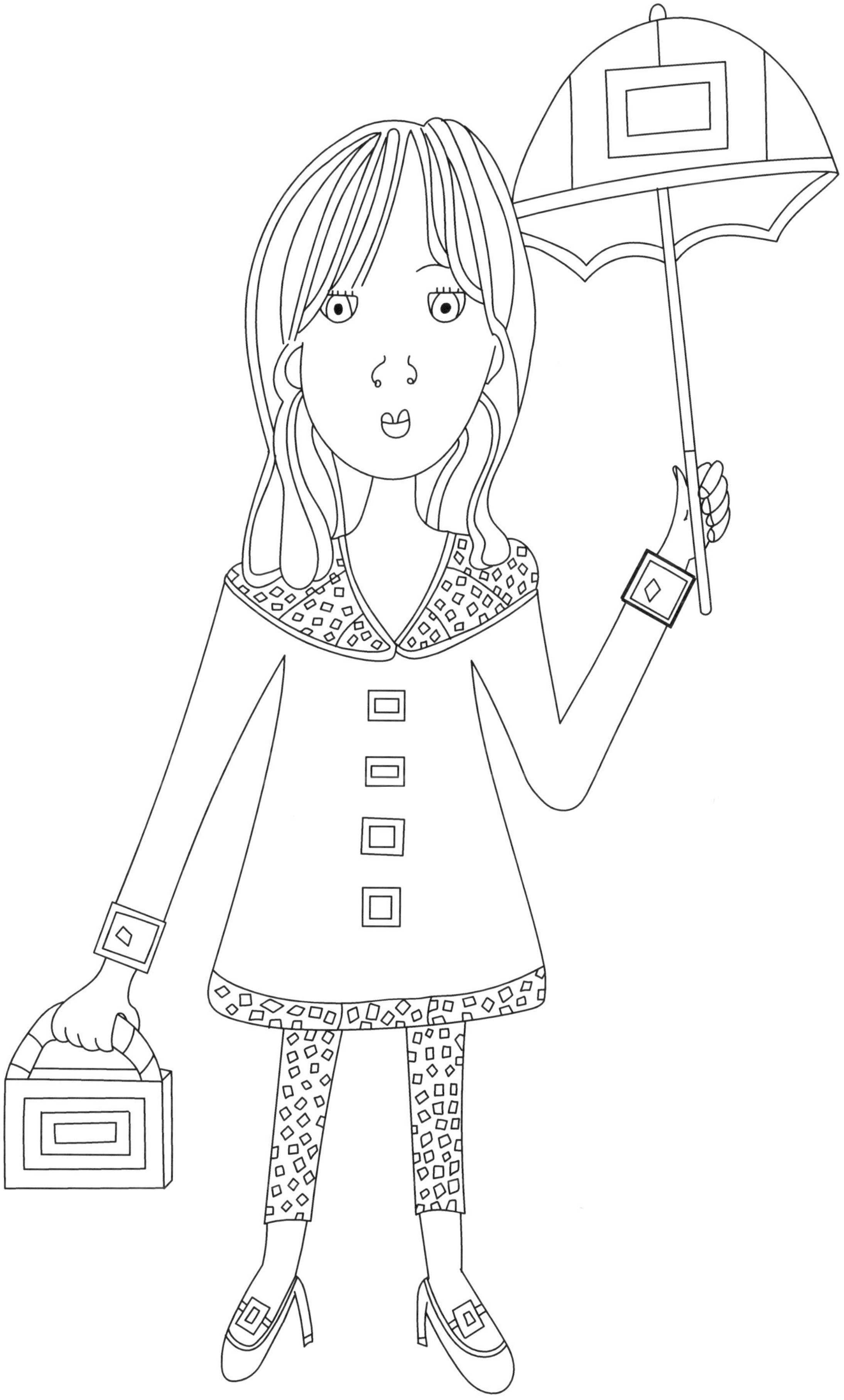

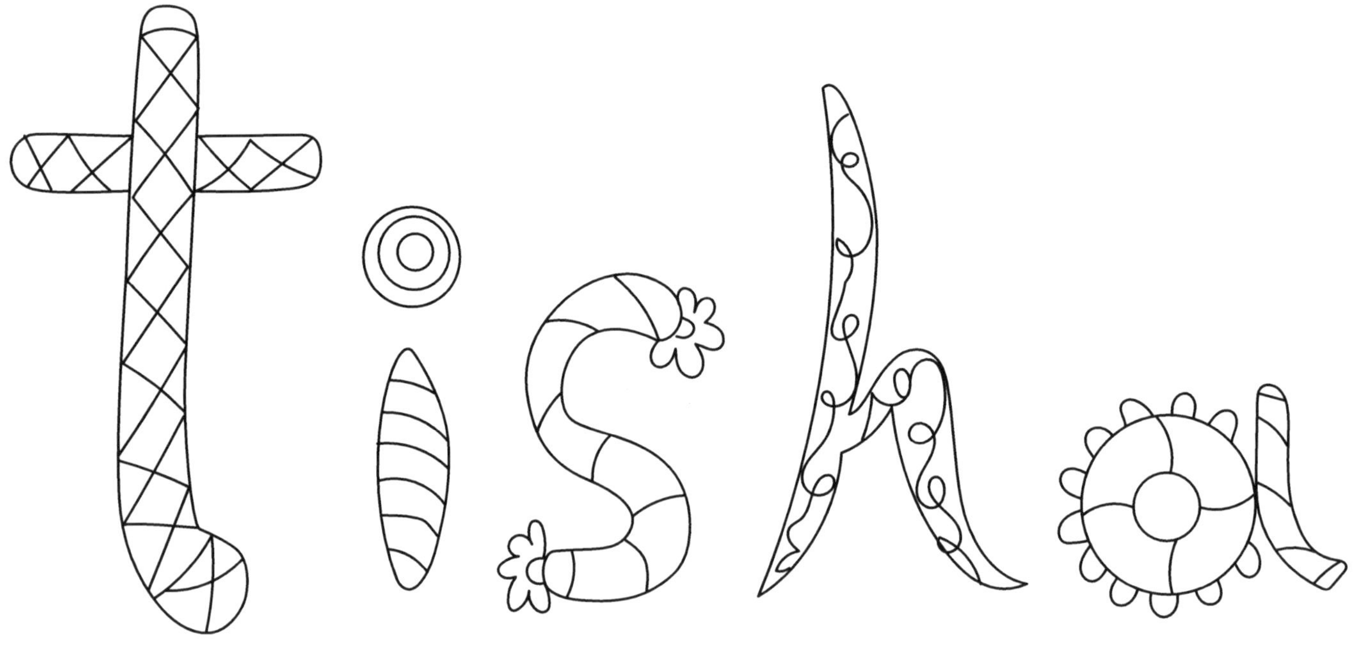
tisha

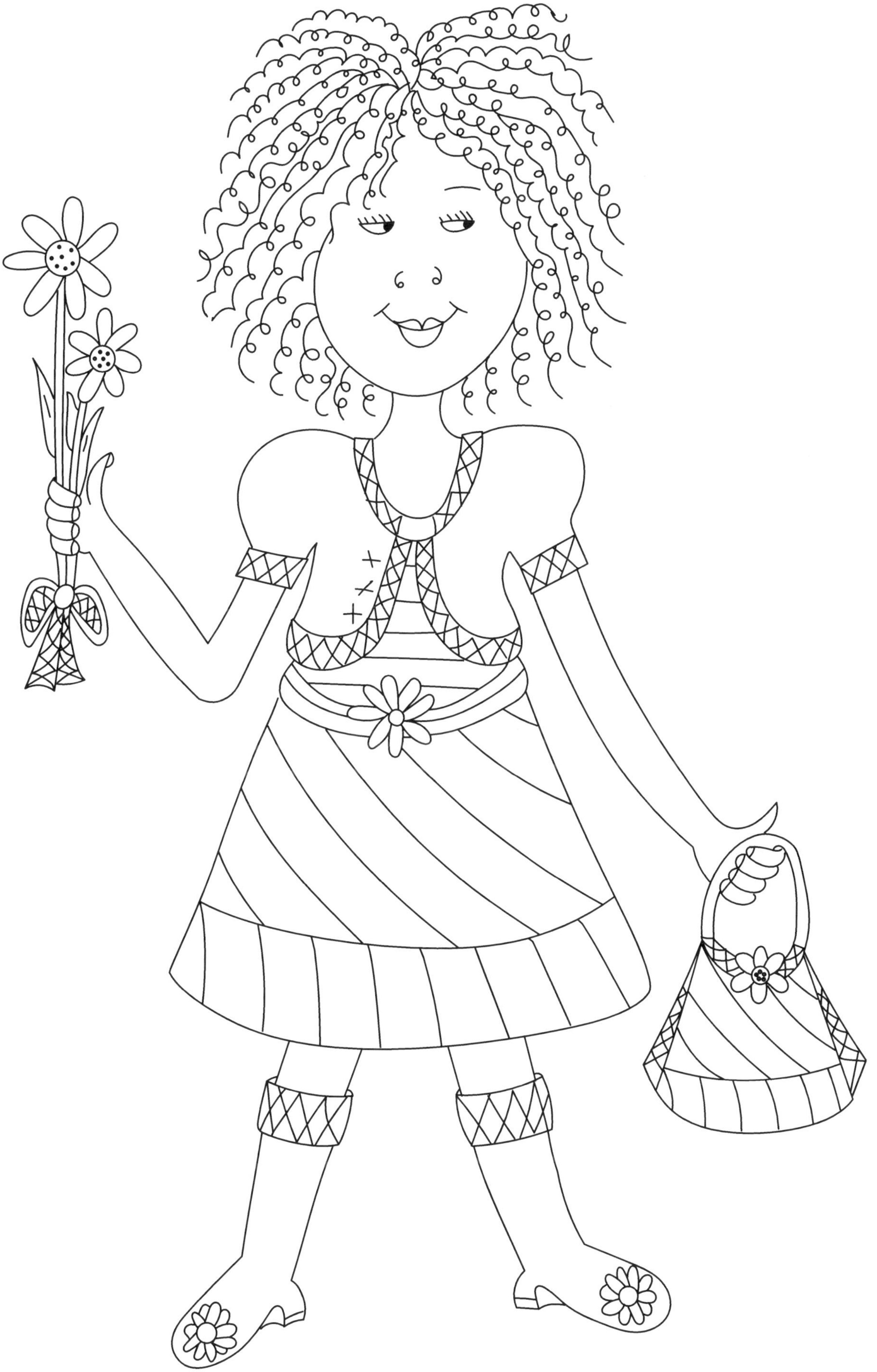

Made in United States
Orlando, FL
04 November 2026

90202922R00022